MPM2D Revision Guide

ONTARIO GRADE 10 ACADEMIC MATHEMATICS

by Mark Burke

ISBN : 1479229563

All Rights Reserved©Mark Burke

HOW TO USE THIS BOOK

This book is a guide to the essential skills that you will need for the grade 10 Ontario academic mathematics course MPM2D.

It is designed to be brief so that it will be both economical and easy for a student to digest. This is is not a comprehensive textbook, and students should practice as many problems as they can from different sources. Ultimately much success in mathematics is down to practice. Nonetheless, all the curriculum expectations have been covered in this booklet, and if students *work* through (rather than just *read* through !) the examples, they should grasp the essential fundamentals of this course.

There is a brief exercise at the end of chapter so that students can test themselves on these fundamentals.

Acknowledgements

Thanks to the students at Astolot Educational Centre for asking the questions that prompted these attempted explanations !

Grateful thanks to Wikipedia for the use of some images in diagrams .

CONTENTS

Unit A : Quadratic Relations.........4
Unit B : Analytic Geometry..........30.
Unit C : Trigonometry................43
Solutions to Review Problems 53

UNIT A - QUADRATIC RELATIONS OF THE FORM $Y = ax^2 + bx + c$

Introduction

A Quadratic Function is an expression with a variable to power 2, such as x^2.

Quadratic Functions can often model a situation that involves an increase, a peak, and then a decrease. In this course we will deal with situations that are modelled with symmetrical curves or parabolas when represented on a graph :

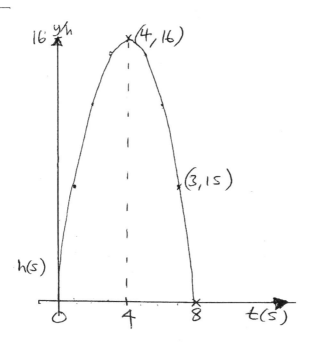

The case above is a classic example. The flight of a model rocket is plotted on a graph where the axis tracks time and the y-axis tracks height.
An equation can be written which might record it's flight. In this case we will use $y = -x^2 + 8x$ or rather:

$h = -t^2 + 8t$ (where t is measured in seconds and height in metres.)

The rocket starts at t = 0 and h = 0, peaks at t = 4, h = 16, and finally lands at t = 8, h = 0.

The quadratic equation rather than just the graph can be used to find any height if given time, for instance if $t = 3$, substitute this into the quadratic equation : $h = -(3)^2 + 8(3) = -9 + 24 = 15$. When $t = 3$, $h = 15$.

A great deal of of this course will be looking at how to find 't' if given h, or how to find x values if given a y-value (usually 0 and the peak or trough) and how to find other information for how the equation looks on a graph.

Most of the equations you will look at are in the **standard form of : $ax^2 + bx + c$**. In this form you can at least derive the y-intercept, where the equation cuts the vertical axis. You also know which way up the graph is ($+x^2$ opens upwards , $-x^2$ is upside down)

You will also be finding the :
-x- intercepts (roots or zeroes
- the axis of symmetry (the vertical mirror line that divides the graph)
- the vertex (highest point or lowest point depending on which way up the graph is) that lies on the axis of symmetry.

On the above example of the rocket, you can see from the *graph* that the axis of symmetry is t = 4, the vertex is (4, 16) and the zeroes are 0 and 8.

Now we need to look at how we derive that information from an *equation* in order to sketch the graph. We could plot a table of values as with any graph but here we want a few key features that enable us to make a quick sketch.

Factorising :

Quadratic functions can be broken down into products so that the zeroes or roots (x - intercepts where the equation cuts the x-axis) can be more easily found. This is the opposite of brackets expansion.

The first thing to do is check for a common factor. For example if

$y = 2x^2 - 8x - 10$ all terms can be divided by 2, so the equation becomes
$y = 2(x^2 - 4x - 5)$

From here we have a variety of strategies to further break this equation into products.

In this particular case we have a **perfect square trinomial** which can be factorised using the factors of the end term (-5) that add in some combination to

make the middle term (- 4).
This is a simple example in which the factors have to be 5 and 1, you just need to make sure you use -5 and 1, NOT 5 and -1, because you want to make NEGATIVE 4 in the middle.

-5 + 1 = -4 (middle coefficient)
-5 x 1 = -5 (end term)

So, $2 (x^2 - 4x - 5) = 2 (x - 5)(x + 1)$

General strategies on factorising trinomials
Notice that the positive and negative signs in the brackets are defined by the coefficients of the middle and end terms :
For $x^2 + bx + c$ you need positive terms in both brackets eg
$(x + r)(x + s)$
For $x^2 + bx - c$ or $x^2 - bx - c$ you need a mix
$(x + r)(x - s)$
For $x^2 - bx + c$ you need negative signs in both
$(x - r)(x - s)$

Factoring perfect trinomials when the co-efficient of x^2 is not 1 :
In the example $y = 2x^2 - 3x - 20$ we cannot take out a common factor, yet the trinomial can still be

factorised as $(2x + a)(x + b)$ where 'a' and 'b' are the factors of -20.

We know we will be dealing with factors of 20 such as 20, 1 and 4, 5 or 2, 10, *but this time one of those factors (b) will be doubled by the 2.*

In this case -4 and 5 will work because the -4 can be doubled to make -8, and $-8 + 5 = -3$ which is the required co-efficient of the middle term.

The equation is factorised as $y = (2x + 5)(x - 4)$.
(Notice how the -4 has to go in the opposite bracket to the 2 in order to be doubled.)

Some expressions can be factorised using the pattern of **difference of two squares. This uses the model $a^2 - b^2 = (a - b)(a + b)$ which can be thought of as : $A - B = (\sqrt{A} + \sqrt{B})(\sqrt{A} - \sqrt{B})$**

If an equation can be written in the form something squared MINUS something else squared then it can be broken into : ' the product of the square root of one times the square root of the other' .

For example :
$y = 4x^2 - 9$ can be factorised into $(2x - 3)(2x + 3)$

Solving Quadratic Equations:
The whole point of factorising is to create an equation

which can be more easily manipulated and in fact solved for when y = 0.

If you have a product, then any value of x which brings one bracket to zero will make the whole equation equal zero.

For example $y = x^2 + 9x + 14$ is factorised as $y = (x+2)(x+7)$

if x = -2, then y = (-2 + 2) (-2 + 7)
     = 0 x 5
     = 0

Likewise if x = -7, then y = (-7 + 2) (-7 + -7)
       = -5 x 0
       = 0

In fact we just have to set each bracket to zero and solve for x.

Another example :

From the earlier problem in 1. 3 we had y = (2x + 5) (x - 4)

Solve 2x + 5 = 0
   2x = - 5
    x = - 2.5

Also solve : x - 4 = 0
     x = 4

The solutions or zeroes / roots of this quadratic are -2.5 and 4.
Remember, graphically this is where the equation intercepts the x-axis.

The Quadratic Formula can sometimes be simplest when solving for the zeros of an equation, and is particularly useful when you have difficult trinomials with multiple possible factors for the end term.

When $x = \dfrac{-b \pm \sqrt{b^2 - 4ac}}{2a}$

the resulting answers will be the zeroes.
Note how because you take the square root of '$b^2 - 4ac$' there will be two answers :

$x = \dfrac{-b + \sqrt{b^2 - 4ac}}{2a}$

and

$x = \dfrac{-b - \sqrt{b^2 - 4ac}}{2a}$

For example consider $y = 3x^2 - 2x - 8$
$a = 3, b = -2$ and $c = -8$

$$x = [-(-2) +/- \sqrt{((-2)^2 - 4 \times 3 \times -8)}] / 2x - 2$$

$$x = [2 +/- \sqrt{4 + 96}] / -4$$

$$x = [2 +/- \sqrt{100}] / -4$$
$$x = (2+10) / -4 \quad \text{OR} \quad x = (2-10) / -4$$

$$x = 12/-4 = \mathbf{-3} \quad \text{OR} \quad x = -8/-4 = \mathbf{2}$$

In fact the part of the quadratic formula under the square root **(the discriminant $b^2 - 4ac$) is also a short cut to finding out how *many* roots there.** Sometimes there may be only one root, and sometimes none.

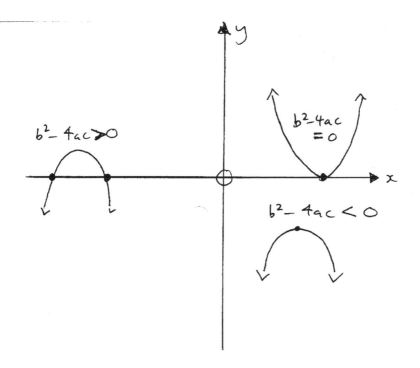

If b − 4ac is more than zero, there are two roots.
If b − 4ac is equal to zero, there is one root.
If b − 4ac is less than zero, there are no roots (in other words on a graph the parabola does not cross the x- axis)

Using the TI-83 or graphical equivalent
Your calculator can also plot a quadratic equation. Enter the equation under Y=, graph, and then use the TRACE function to locate the x-values when y =0.

Rewriting in Vertex form and Completing the square

This is a method to isolate the x variable so it can be written in vertex form, which can also be solved to find the zeroes or x -intercepts.
In fact if an equation is written in the form $y = a(x - h)^2 + k$ the **vertex** can be located immediately as **(h, k)**
For example the vertex in the equation $y = -3(x + 4)^2 - 5$ is
$(-4, -5)$.
For this reason we call $a(x - h)^2 + k$ **the vertex form.**

Consider the equation $y = x^2 + 4x - 11$ which at the moment is in standard form.
We start by using $(x + b/2)^2$ to produce the first two terms (in other words use half of the middle term)
In this example $x^2 + 4x - 11$ is rewritten as $(x + 2)^2$.
Whilst $(x + 2)(x + 2)$ does produce $x^2 + 4x$ it also produces 4 as an end term.
Our standard form has -11, not 4 as an end term, so to 'complete the square we have to subtract 15. Subtracting 15 will get us from 4 to -11.
Thus, $x^2 + 4x - 11 = (x + 2)^2 - 15$.

The equation is now in vertex form, where vertex is (-2, -15)

Vertex form can also be solved to find the x-intercepts. Set the equation to zero and rearrange for x.

If $(x + 2)^2 - 15 = 0$
$(x + 2)^2 = 15$
$x + 2 = +/- \sqrt{15}$
(square root both sides)
$x = +/- \sqrt{15} - 2$

Note how there are two answers ;
$x = + \sqrt{15} - 2 = 1.8$ AND
$x = - \sqrt{15} - 2 = -5.87$

Completing the square and Vertex form when a > 1

If an equation has a coefficient of a that is more than 1, you must factor this out from the first two terms *before* you complete the square. Then once the square is completed, expand the brackets.

For example if $y = 2x^2 - 6x - 19$

Factor out the 2 from the first and middle terms :

$$y = 2(x^2 - 3x) - 19$$

complete the square inside the brackets and ignore the -19 for now.
$$y = 2\{(x - 1.5)^2 - 1.5^2\} - 19$$

(subtract the extra middle term squared, as if you wanted last term 0.)
$$y = 2\{(x - 1.5)^2 - 2.25\} - 19$$

Now expand : $y = 2(x - 1.5)^2 - 4.5 - 19$

Now at the last stage combine the end terms to create vertex form :

$2(x - 1.5)^2 - 23.5$ ie. Vertex is $(1.5, -23.5)$
Set to zero and solve as before to find x-intercepts.

Completing the Square and the Quadratic Formula.

Incidentally, the quadratic formula comes from completing the square on the general standard form $ax^2 + bx + c$.

CHALLENGE : See if you can create the quadratic formula ;
Hint : Start by factoring out a :
$a(x^2 + b/a \, x) + c = 0$
Then use half of b/a : $a[(x + b/2a)^2 - ?] + c = 0$

Now as before, complete the square and solve for x.

- You can now apply all of the theory you have studied to solve more practical problems which are modelled with quadratic functions.

Quadratic equations can also model other situations, such as :

A picture of dimensions 25 cm by 20cm is framed with unknown dimensions, except that the frame is the same thickness all the way through. With framing, the total area of the picture is 600cm². Find the dimensions of the frame.

Use the dimensions (20 + 2x) and (25 + 2x)

so $\quad (2x + 20)(2x + 25) = 1000$
Expand : $\quad 4x^2 + 50x + 40x + 500 = 1000$
Set to zero : $\quad 4x^2 + 95x - 500 = 0$
Solve with a method of your choice to find x.

At this point we should discuss further the **axis of symmetry and types of vertex**. If the parabola is opening upwards (increasing from the mirror line or **axis of symmetry**) then the vertex is a the lowest point of the graph and is called the **minimum.**

If the vertex opens downwards then the turning point is the highest point and is called a **maximum**.

Locating the axis of symmetry (the vertical line that cuts the graph into a left and right half) can be done by using a simple formula
 x= -b /2a   (where b is the coefficient of x, and a the coefficient of x^2).

For example in the equation $y = 2x^2 - 3x + 5$
  a = 2 and b = - 3 , so the axis of symmetry is :   x = - (-3) / 2x2 = 3/4.

The vertex is on the axis of symmetry and you simply input your answer for x into the function to find the corresponding y -value.
In the above example,   if x = 3/4, then y = 2 (3/4) ^{2} - 3(3/4) + 5 = 3.875
So,   the point (0.75, 3.875) is the vertex (in this case a minimum because the graph opens upwards)

The axis of symmetry is also halfway between the roots and is especially easy to locate if the equation is in :
Factored form   y = a (x- r) (x- s)
In this factored form, r and s are the zeroes or x- intercepts. For example in the equation

$y = 3(x-2)(x+4)$
the roots or zeroes are 2 and -4.

The x – intercepts, when they exist, can also be used to locate the vertex.
Remember the vertex is exactly halfway between the zeroes. In the above example, if the roots are 2 and – 4, then the gap between them is 6 units. 3 units in from either point will bring you to $x = -1$.
if the vertex is at $x = -1$, then the corresponding y value is :

$y = 3(-1-2)(-1+4) = 3(-3)(3) = -27.$

Both factored form and vertex form can be returned to standard form by expanding the brackets and simplifying.

The above example expands to $y = 3(x-2)(x+4)$
$= 3(x^2 + 4x - 2x - 8)$
$= 3(x^2 + 2x - 8)$
$= 3x^2 + 6x - 24$

<u>Finding an equation from a graph</u>

If given a graphed equation you can use the vertex and zeroes and one other point to derive a formula which can then be solved for 'a', the steepness of the graph. Together this will make an equation of the graph

For example, if given that the zeroes on a graph are 4 and -3, and the equation passes through (1, 8), derive the equation :

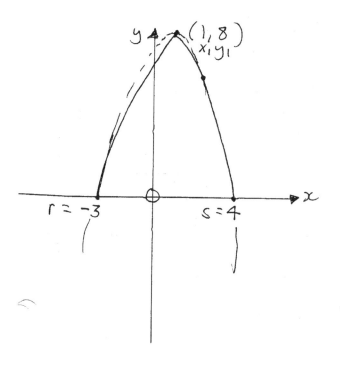

Use $y = a(x - r)(x - s)$
In this case $y = a(x - 4)(x + 3)$
Now we just sub in a an x and y value and solve for 'a'.
Using (1, 8)

$8 = a(1-4)(1+3)$
$8 = a(-3)(4)$
$8 = -12a$
$a = 8/-12$
$a = 2/-3$

Thus the equation that fits the graph is $y = (2/-3)(x-4)(x+3)$
which in standard form is $y = (2/-3)(x^2 - x - 12)$
$y = (2/-3)x^2 + (2/3)x + 8$

A similar procedure can be done with vertex form ; if given a vertex such as (-2, 6) and the point (2, 1) then use $y = a(x-h)^2 + k$
Here $1 = a(2 - -2)^2 + 6$
$1 = 16a + 6$
$16a = -5$
$a = -5/16$
Thus the equation that fist the graph is
$y = (-5/16)(x+2)^2 + 6$

Transformations

It is possible to sketch accurate graphs of equation by transforming the 'base function'.
If for example the base function is x^2 then the function $-3(x+4)^2 - 5$

can be seen a series of transformations of $y = x^2$
In the general form the transformation is labelled $y = a(x-h)^2 + k$
where

a = a vertical stretch, ie the graph gets steeper
(If a is negative, the graph is reflected in the x-axis.)
h = a horizontal shift where the graph is moved 'h' units right
k = a vertical shift that moves the graph up by 'k' units.

Next it is important that you apply the transformations in the right order.
i) Stretch and reflect the graph before you shift it.
ii) The horizontal and vertical shifts can be done in any order.

In our example above the graph can be transformed in 3 stages.
- it is stretched to become 3 times steeper. The y-values become 3 times bigger)
- it is reflected in the x-axis, effectively being flipped upside down.
- it shifted 4 units left . (note how + 4 moves the graph *left*).
- it is shifted 5 units down.
-

You can see the effect in a table of values:

x	x^2	$3x^2$	$-3x^2$	$-3(x+4)^2$	$-3(x+4)^2 - 5$
0	0	0	0	-48	-53
1	1	3	-3	-75	-80
2	4	13	-12	-108	-113
3	9	27	-27	-147	-152

-Or graphically : (the base function x^2 is graphed on each faintly)

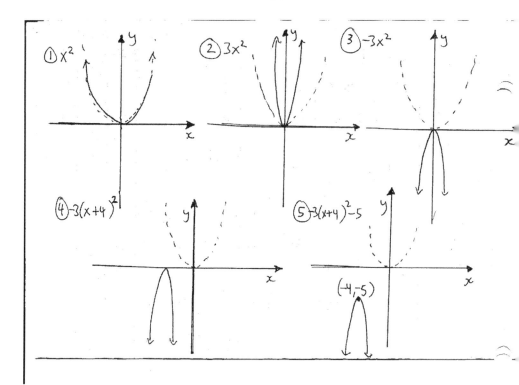

Comparing Qaudratic Relations to Exponential Relations

You can tell if a relation is quadratic by looking at the second differences. Consider for example the relation $y = x^2$.

x	0	1	2	3
y	0	1	4	9

The first differences in the output/y values are 1, 3, and 5.
The differences of *those* differences are 2, 2, etc. Because this is constant, the relation is quadratic.

We can identify exponentials ($y = a^x$) by the same method. An exponential is a relation where a base is to the power x. If the first difference reveals a multiplicative difference (e.g the differences are doubling or trebling etc) then the relation is exponential.
Consider the function $y = 2^x$;

x	0	1	2	3	4	5
y	1	2	4	8	16	32

You can see the output/y values are doubling, so this is an exponential function.

Graphically, exponentials climb at an accelerating rate and look like this ;

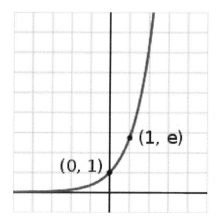

(The above example is a very special function where the power is 'e' , about 2.718. You will work with this much more in grade 11 for now think of this as y = 2.719^x)

Negative exponents will produce fractions, because a negative exponent is a reciprocal that means 1/a.
2^{-3} for instance means $1 / 2^3$ = 1/8

Grapically, whereas a positive exponential will climb at an accelerating rate, a negative exponential will get smaller and smaller and approach the x-axis.

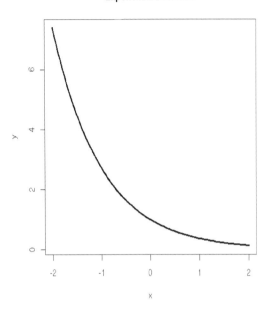

Exponential Decrease

You should also know the basic rules of exponentials ;

1) A base to the power of zero is 1
   $x^0 = 1$

2) A base to the power of 1 is itself
   $x^1 = x$

3) A base to power 1/n is to the nth root.
   eg. $16^{\frac{1}{2}} = 4$ ($4^2 = 16$)

4) A base to a negative power is a reciprocal

$$X^{-3} = 1/x^3$$
$$\text{e.g } 4^{-2} = 1/4^2 = 1/16$$

5) A fractional exponent is where the denominator is the root and the numerator the power. e.g

$$16^{\frac{3}{4}} = (16^{\frac{1}{4}})^3 = 2^3 = 8$$

6) A base to a power itself raised to a power multiplies the powers
e.g, $(8^2)^{\frac{1}{4}} = 8^{\frac{1}{2}}$

7) A base to raised to a power multiplied by the same <u>base</u> adds the powers eg.
$8^1 \times 8^2 = 8^3$

8) A base to a power divided by the same base subtracts the powers
e.g $2^{\frac{3}{4}} / 2^{\frac{1}{4}} = 2^{\frac{1}{2}}$

<u>Mixed example</u> : It is generally best to work out negative exponents first, then roots, then powers
e.g :
$16^{-\frac{3}{4}} = 1/16^{\frac{3}{4}}$
$1/(16^{\frac{1}{4}})^3 = 1/2^3 = 1/8$

UNIT A REVIEW PROBLEM (see back of booklet for solutions)

1) A profit and spending function is modelled by $P = -2s^2 + 10s - 1$ where s and P are in thousands of dollars.

i) Find what level of spending maximizes profit, and how much that profit is.

ii) Find what level of spending brings profit back to zero.

iii) Find profit when spending is 3000 dollars.

iv) graph the equation and clearly show y-intercept, axis of symmetry, vertex and zeroes.

v) Calculate $8^{(-2/3)}$

UNIT B : ANALYTIC GEOMETRY

Review :
$y = mx + b$: the general equation of a line

When plotting equations, you can find the gradient or **slope of a line** by looking at the number in front of x *if the equation is in the form $y = mx + b$.*

In the above example, if $y = \underline{3}x - 4$, then the slope is $+\underline{3}$. For every one unit you go across, you go up 3 units.

If an equation is not in the form $y = mx + b$ then you can rearrange it so it is.
E.g if given $2y - 4x = 6$, rearrange for y :

$2y = 6 + 4x$
$y = 3 + \underline{2}x$. Now you can tell that the slope is 2 (the number if front of x).

The general equation of a line also tells you the y-intercept ; the point at which the line crosses the vertical axis. In the equation this is the constant ; that is the number on it's own
 In the equation $y = 3x - 4$, the y-intercept is - 4.

Finding intersections of lines using graphical and algebraic methods

a) <u>Graphical Method</u>:

Plot each equation by using a table of values or other

e.g : Solve i) $y = 3x + 1$, and ii) $y = 2x - 4$.

For $y = 3x + 1$, try $x = 0$;

- If $x = 0, y = 3(0) + 1 = 1$. So we have the point (0, 1).

Repeat for another x-value of $y = 3x + 1$:

- If $x = 1, y = 3(1) + 1 = 4$. Hence we have the point (1, 4).

Plot both these points and join them up to draw the line of $y = 3x + 1$.

It should look like the diagram below.

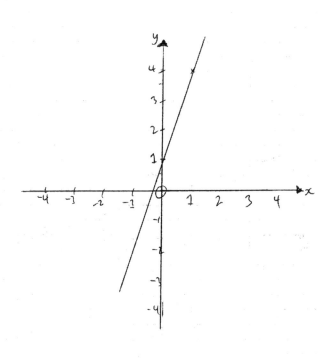

For y = 2x − 4, first try x = 0 :

If x = 0 , y = 2(0) - 4 = -4. So we have the point (0, -4).

If x = 1, y = 2 (1) − 4 = -2. So we have the second point (1, -2)

- Plot this line on the same graph :

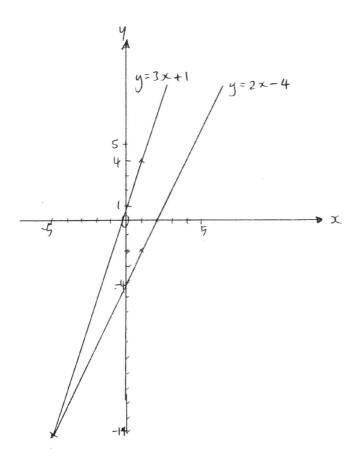

The solution is where the 2 lines intersect ; in this case (-5, -14).

At this particular point both equations have the same x and y values.

b) Algebraic Substitution:

If the equations are not in the form $y = mx + b$, it may be more convenient to rearrange one of them and sub it in to the other equation.

e.g i) $x + y = 3$ and ii) $2y + 4x = 7$.

It's easiest to rearrange (i); If $x + y = 3$, then $y = 3 - x$

Replace 'y' in the second equation with '$3 - x$'.

$$2(3 - x) + 4x = 7$$

$$6 - 2x + 4x = 7$$

$$2x = 7 - 6$$

$$x = 1/2$$

Now take the first solution and sub it in to the first equation :

i. If $y = 3 - x$,

y = 3 - 1/2

y = 2½

The solution is (1/2, 2½)

c) Algebraic Elimination :

Sometimes equations can be lined up and added or subtracted to eliminate a variable. It may be necessary to multiply both equations by a number to get the same number of a certain variable. You want the same number of x's or y's in both equations, so that when you add or subtract equations one of the variables does actually get eliminated.

eg.

For the equations i) $2x - 3y = -6$ and ii) $3x - 5y = -11$

Multiply the first equation by 3, and the second equation by 2, so you have *6x* in both equations. Note that everything must be multiplied by the 2 and 3.

x3 i) $6x - 9y = -18$

x2 ii) $6x - 10y = -22$

Because we have 6x in both equations we can subtract one from the other to eliminate x and solve for y :

i)- ii) : (6x - 6x) + (-9y - - 10y) = - 18 - - 22

0 - 9y +10y = -18 + 22

y = 4

Now go back to one of the *original* equations and solve for the other variable.

If y = 4, i) 2 x - 3 (4) = - 6

2x - 12 = - 6

2x = 6

x = 3.

The full solution is (3, 4).

Application of Linear systems to Word Problems

In most of these problems, x and y represent quantities of a variable and will be expressed as ;

i. x + y = total quantity

ii. x (cost of 1 x) + y (cost of 1y) = total cost

e.g. A large group of people buy tickets at a cinema. A total of 32 tickets are bought ; a mix of adult and child tickets. Child tickets cost 8 dollars, and adult tickets 10 dollars. The total cost of all the tickets comes to 280 dollars. How many of each type of ticket were bought ?

Equation 1 expresses the total quantity : i) x + y = 32

Equation 2 shows total cost : ii) 8x + 10 y = 280

This can now be solved using a method of your choice.

Properties of line segments

We can find the <u>midpoint of a line</u> segment simply by averaging the co-ordinates. e.g the midpoint of the line segment joined by (3, 8) and (7. 18) is (3+7/2 , 8 +18/2) = (5 , 13)

The <u>length of a line</u> can be calculated by using pythagoras. The line is the hypotenuse of a right triangle where you have to find the height and length.

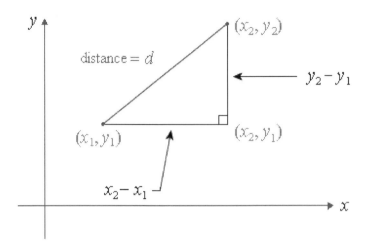

Note how (x^1, y^1) is always the point on the left

Consider the line segment that joins (-2, 5) and (4, -6). The difference between the y values (height) is 4 - (-2) = 6. The difference between the y -values is -6-5 = -11

Pythagoras tell us the hypotenuse is $\sqrt{(-11)^2 + (6)^2}$

$\sqrt{(121+36)} = \sqrt{(157)} = 12.5$

The <u>slope of a line</u> is calculated by calculating the difference in the y co-ordinates and *dividing* by the difference between x co-ordinates.

For example the slope of the line that joins (1, - 4) and (4, - 18) is (-18--4)/ (4 - 1) = -14/3

The <u>Perpendicular Bisector</u> of a line is the line that cuts the first line in half at an angle of 90°.

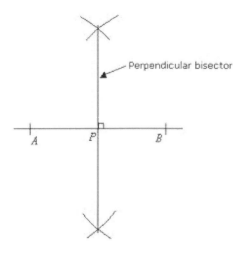

The first thing to find is the half way point by using midpoint of a line.

Secondly you find slope of the line 'm' and convert it into a reciprocal using 1/-m. e.g the slope of a line that is 2 has a slope of $-\frac{1}{2}$.

(Perpendicular slopes make a product of -1 when multiplied together.)

Combine these to make an equation of the new line by

inputting m,x and y to find b in y = mx + b

Example : the perpendicular bisector of the line that joins (2, -3) and (4, 8) :

The midpoint is (3 , 2.5).

The slope is (11/2) = 5.5

The equation of the perpendicular bisector is using y = mx + b where m = 5.5 and (3, 2.5) is x and y

$$2.5 = 5.5(3) + b$$
$$2.5 = 16.5 + b$$
$$b = 2.5 - 16.5$$
$$b = -14$$

The Equation of the perpendicular bisector is y = 5.5x -14

Equations of Circles

The general equation of a circle centred on (0,0) is : $x^2 + y^2 = r^2$, where r = radius. This formula comes from the pythagoras formula for length of a line.

For example $x^2 + y^2 = 5^2$ is the circle that sits on the origin (0,0) and has a radius of 5. It goes through (0,5) (5, 0) (-5, 0) and (0, -5)

Usually you have to square root the constant to find radius. e.g

$x^2 + y^2 = 36$ has a radius of 6.

Using analytic geometry to verify properties of shapes

You can use the theory you have just learned to confirm that points on graphs do really make certain shapes. For example we can use perpendicular bisectors and lengths of lines to prove that four points make a rectangle.

Example : Prove that the points (1,1) (1,3) (4,3) and (4, 5) form a rectangle.

We need to prove that some of the lines are perpendicular and that the parallel lines are equal in length.

See if you can prove

i) the lines that join (1,3) and (4,5) are parallel to the lines that join (1,1) to (4,3) and then prove

ii) that these lines are perpendicular to the lines that join (1,3) to (1,1) and (4,3) to (4,5). Here use the fact that if you multiply the slopes the product is -1.

and then prove

iii) the parallel lines are equal in length

All of this confirms the shape is indeed a rectangle.

UNIT B REVIEW

1) Find the slope and y-intercept :

  i) $2y + 7 = x$ ii) $1/2\ y - x = 4$

2) Find the intersection of the lines $y = 3x - 1$ and

  $y = -4x + 2$

3) Find the perpendicular bisector of the line that joins (2, 6) and (5, 9)

4) Draw a rectangle on squared paper and *prove by geometric properties of lines* that is is a rectangle

UNIT C - TRIGONOMETRY

Similar Triangles

If triangles are similar it means that they are essentially bigger or smaller versions of each other and the angles inside are the same.
A simple ratio and cross multiplying will solve most problems. The key is to make sure you have compared like sides by checking the positions of the angles.

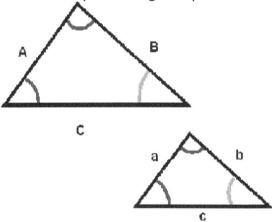

In this triangle A/a = B/b = C/c
If we had lengths, such as A = 10, B = 12 and a = 7, we can solve the like ratios for a missing side such as 'b'.

10/7 = 12 / b

switch everything upside down to make it easier to find x :

7/10 = b /12
(7/10) (12) = b
$\quad\quad$ 8.4= b

Many problems involve a right triangle inside another :

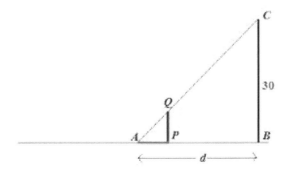

You can see here angle QAP is equal to CAB and both triangles are right triangles, so they are similar.

Testing for similarity
By comparing the ratios of sides you can confirm similarity.
If the above triangle had sides A = 15 , B = 20, is the triangle with sides a = 6, b = 8 similiar ?
Since 15/20 = 3/4 and
$\quad\quad$ 6/8 = 3/4 then the ratio of like sides is the same and therefore the triangles are indeed similar.

Congruency is similarity where the triangles are exactly the same *size* as well having the same angles

Trigonometry in Right triangles using Sine (Sin.), Cosine (Cos.) and Tangent (Tan.) ratios

-There are 3 formulae that can be used to find angles inside right triangles, and find lengths if given an angle and one other length.
-They can be remembered as 'SOHCAHTOA' (*Saw-Ka-Toe-Ah*)
Where sin x = **o**pposite/ **h**ypotenuse, cos x = **a**djacent/ hypotenuse, and tan x = opposite / adjacent .

First of all, you must understand that although the hypotenuse is always the same in a right triangle (longest side, opposite the right angle), *the adjacent and opposite can change depending on what angle you are working with.*

In the two congruent triangles below, the adjacent and opposite switch around depending on whether you use angle 'x' or angle 'y'. An easy way to remember this is that the adjacent is always the shorter side touching the angle, whilst the opposite is literally across from the angle.

Finding angles if given two sides

-First label the sides of the triangle according to which angle you are trying to find.
E.g in this case the sides from angle 'x' are hypotenuse and adjacent.

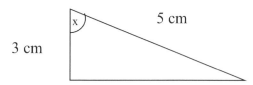

-Next choose the appropriate formula ; in this case cos x = adjacent/ hypotenuse.

—   Write out the ratio, and use the inverse/ shift function on your calculator on that ratio. In this case :

Cos x° = 3/ 5
cos $^{-1}$ (3/ 5) = 53.1 °

Finding a side if given an angle and another side

As before, write out the ratio based on what sides are involved *relative to the angle.*
e.g. In this case because we have the adjacent and we

are looking for the opposite (relative to the angle), we use the tan ratio :

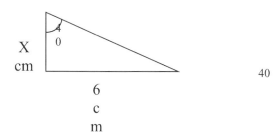

40° = adj. / 6cm

tan 40° x 6 = adjacent
x = 5cm

- Beware of getting the ratio the right way up and the trickier algebra if the side you are looking for is on the denominator. Eg.2 ;

In this triangle we have the opposite and need the hypotenuse, so we use Sine.

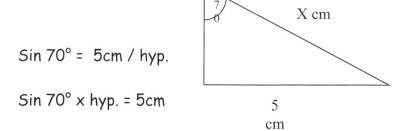

Sin 70° = 5cm / hyp.

Sin 70° x hyp. = 5cm

hypotenuse = 5cm / (Sin 70°) = 5.3 cm

Trigonometry in Non Right Triangles

The Sine Law SinA/a = SinB/ b and the
Cosine Law a^2 = b + c -2bc Cos A
can be used in non right triangles where you have a corresponding angle and side. By corresponding here, we mean an angle corresponds to its **opposite side**.

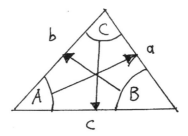

Each law is best used in certain situations.
If you have a problem involving two pairs of angles/sides, use **Sine Law**. The law can be written upside down if calculating sides, for example in the problem below we have used a / SinA = b / Sin B :

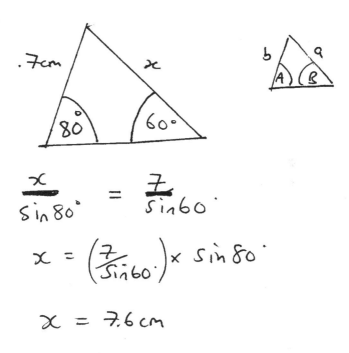

$$\frac{x}{\sin 80°} = \frac{7}{\sin 60°}$$

$$x = \left(\frac{7}{\sin 60°}\right) \times \sin 80°$$

$$x = 7.6 \text{ cm}$$

Cosine Law is used in situations in which you have two sides and the 'included' angle between them, OR you have all 3 sides and NO angles.
For example in the problem below :

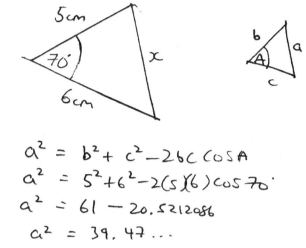

$$a^2 = b^2 + c^2 - 2bc \cos A$$
$$a^2 = 5^2 + 6^2 - 2(5)(6)\cos 70°$$
$$a^2 = 61 - 20.5212086$$
$$a^2 = 39.47...$$
$$a = 6.28 \text{ cm}$$

In this example we have 3 sides and find an angle :

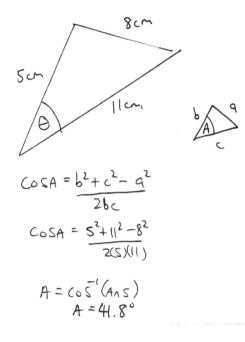

$$\cos A = \frac{b^2 + c^2 - a^2}{2bc}$$

$$\cos A = \frac{5^2 + 11^2 - 8^2}{2(5)(11)}$$

$$A = \cos^{-1}(\text{Ans})$$
$$A = 41.8°$$

UNIT C REVIEW

i) Calculate the angle x.

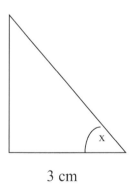

5 cm

3 cm

i)   Calculate the side x

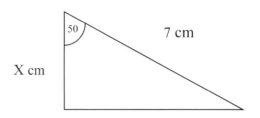

2) Solve (find all missing values) in the triangles ;

(i)

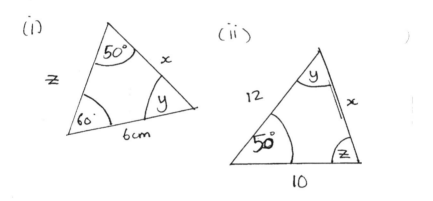

(ii)

3) If A = 6 and B = 9, find b in the similar triangle where a = 5 .

UNIT REVIEW SOLUTIONS

UNIT A REVIEW PROBLEMS SOLUTIONS

i) s = 2.5 , P = 11.5 , so spending $2,500 maximises profit at 11,500
i) s = 4.9
ii) P = 11

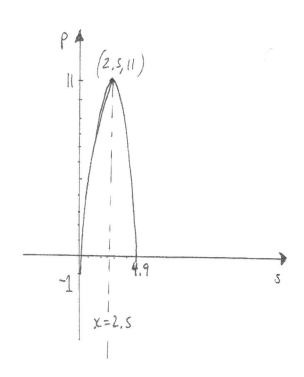

iv) see above
v) 8^(-2/3) = 1/8^(2/3) = 1/ 2² = 1/4

UNIT B REVIEW PROBLEMS : SOLUTIONS

1) i) m=1/2 , b = -3.5 ii) m=1, b =8
2) x = 3/7, y = 2/7
3) midpoint is (3.5, 7.5) , m= -1
   so y = mx + b
   7.5 = (-1)(3.5) + b
   11 = b
   So equation of perpendicular bisector is :
   y = -x + 11
4) Answers will vary

UNIT C REVIEW PROBLEMS : SOLUTIONS

1) i) X = 59°   ii) x = 4.5
2) x = 6.78, Y = 70°, z = 7.36
ii) x^2 = 89.73 ; x = 9.47.   Z = 76° , Y = 54°
3) 6/9 = 5/ x ;   x = 7.5

Manufactured by Amazon.ca
Bolton, ON